Piper,

Private Eye

by Karen Tayleur

illustrated by Brad Quinn

EDUCATORS PUBLISHING SERVICE
Cambridge and Toronto

Series Authors: Kay Kovalevs and Alison Dewsbury
Series Consultant: Melinda Rice
Commissioning Editors: Sarah Russell, Bonnie Lass, and
Sethany Rancier Alongi
Senior Editorial Manager: Sheila Neylon
Text by Karen Tayleur
Illustrated by Brad Quinn
Edited by Beth Browne, Elissa Gershowitz, and Laura Woollett
Designed by Jennifer Johnston

Making Connections program developed by Educators Publishing
Service, a division of Delta Education, LLC and by Harcourt
Education, a division of Reed International Books Australia Pty Ltd

ISBN 0-8388-3328-4

2010 2009 2008 2007 2006
10 9 8 7 6 5 4 3 2 1

Printed in China

Contents

Chapter 1

Detective Dreams

THERE WAS ONLY ONE TV show Caleb Piper liked. He couldn't tell anyone about it, though, because it was kind of uncool. The program was actually for younger kids. It was a cartoon called *Detective Magnet, Private Eye*. If the other kids at school knew he liked *Detective Magnet*, they'd laugh him out of town.

Caleb daydreamed about being a private eye. Sometimes, he thought about it when he was supposed to be doing other things, like listening to Mr. Murray, his teacher.

"Caleb Piper, are you paying attention?" Mr. Murray asked at least twice a week. "What is so interesting out that window?"

But Caleb wasn't looking at anything out the window. He couldn't tell his teacher the truth: that he was actually picturing himself as a private detective. Caleb didn't think Mr. Murray would understand. He would probably think Caleb was too young to be a detective. But Caleb knew he wasn't too young. If only there was a mystery to solve, Caleb could show everyone.

Unfortunately, life was fairly boring in Caleb's neighborhood, and the apartment building where he lived. It *was* boring . . . until Tyson Jacobs came along.

“There’s a new kid moving into the apartment next door,” said Caleb’s dad one day. “You should get to know him. Who knows? You might just end up being the best of friends.”

Caleb shrugged. He wasn’t sure. He was sometimes shy when it came to meeting new people.

Then he met Tyson.

Chapter 2

Eye Spy

TYSON JACOBS was unlike any other kid Caleb knew. The first time they met, Caleb was hunting for his lost baseball in the yard at the back of the apartment building. He was thinking: Where would Detective Magnet look?

"It's under the clump of grass," said a voice.

Caleb glanced behind him. There was no one there. He spun around in a circle, but still didn't see anyone.

"The clump of grass," the voice repeated.

Caleb looked under the clump of grass. Sure enough, his ball was half hidden in the soil. He grabbed the ball and turned around again.

“Aren’t you going to thank me?” asked the voice. “I’m up here.”

Caleb peered up at the large oak tree that shaded his face. In one fork of the tree sat a boy.

“Hi, I’m Tyson,” said the boy. “You must be Caleb.”

Caleb nodded. He wasn’t sure whether to be angry or not that someone had been secretly watching him search for his ball. He wondered if he’d done anything embarrassing.

“This yard is pretty cool,” Tyson commented. “At the last apartment building where we lived, there was no yard. I used to watch tons of television. I even got into *Detective Magnet*.”

Caleb forgot about being mad. "You like *Detective Magnet*?" he asked.

"Yeah," said Tyson, looking slightly embarrassed. "Got a problem with that?"

That's when Caleb knew he and Tyson were going to be good friends.

Later that afternoon, Caleb and Tyson watched *Detective Magnet* in Tyson's apartment. There were boxes everywhere. There was dust on every surface and clothing hanging in doorways.

"We haven't finished unpacking yet," explained Tyson. "Our couches haven't even arrived."

Tyson was sitting on a large box marked "kitchen." He was resting his feet on a smaller box marked "books." The boys had raided the kitchen. Caleb reached for another pretzel and started crunching.

"Ssshhh," said Tyson. "It's back on."

This episode was a tough case, as the voice-over explained: "The president of a toy company is missing. His car is still in its garage. His mail is piled up in his mailbox. He didn't call his mother on her birthday. Everyone is worried about him. There is only one person they can call . . ."

"Detective Magnet!" exclaimed Tyson and Caleb together.

Detective Magnet visited the toy factory, and the workers told him what they knew. "Hmm . . . I think we have a mystery on our hands," said Detective Magnet.

There were many twists and turns in the plot. Large toy trucks chased after Detective Magnet's car. Robots tried to scare him away. But Detective Magnet kept investigating.

"I'll bet it's a nasty villain trying to take over the world," guessed Caleb.

"No, it's some greedy person who wants to take over the toy factory," said Tyson.

In fact, it was both: a villain wanted to use the toy factory to create monster toys to help him take over the world. When Detective Magnet rescued the president from the toy factory's basement, Caleb and Tyson cheered, "Go, Detective Magnet!"

"Okay, outside, you two. Time for some fresh air," said Mrs. Jacobs.

Caleb and Tyson headed outside and started looking for clues.

"A broken twig," pointed out Tyson.

"A footprint," said Caleb, leaning close to the flowerbed.

"So many clues!" said Tyson. "If only there were a mystery to solve."

"Nothing ever happens around here," agreed Caleb. "Anyway, I've got to go. It's dinner time." The boys went inside the lobby to take the elevator to the fourteenth floor.

"Hang on, I'll go check the mail. My dad always forgets to get it," Caleb said. When the boys approached the mailboxes, Caleb noticed something. The morning newspaper and mail were still in Mr. González's box. Mr. González lived on the same floor as Caleb and Tyson. He always picked up his mail promptly. That's strange, thought Caleb.

The next morning when he was on his way to school, Caleb looked in at Mr. González's mailbox. Now there were two newspapers. When he came home from school, the newspapers still hadn't moved and there was a large pile of mail.

"Hmm," said Caleb. "I think we have a mystery on our hands."

Chapter 3

A Neighborhood Mystery

AFTER FINISHING his after-school chores, Caleb knocked on Tyson's door.

"We have a neighborhood mystery to solve!" he exclaimed. Caleb began to explain about the unopened newspaper in Mr. González's mailbox.

"So he forgot to get his newspaper one morning," interrupted Tyson with a shrug. "So what?"

Caleb went on to explain about the second newspaper and the mail.

"Maybe he's just messy," said Tyson.

Caleb shook his head. Mr. González had the tidiest apartment in the whole building.

"I think we should go check it out," said Caleb.

"But what can we do?" offered Tyson.

"We'll just knock on his door," said Caleb, "to see if he's home."

"And if he opens the door?" asked Tyson.

"We'll think of something," replied Caleb.

As Tyson and Caleb started to leave, Tyson's mother asked them to deliver a letter to Mr. González. One of his letters had been delivered to their mailbox by mistake.

"Could you just take it over to number 14B?" asked Mrs. Jacobs.

Tyson and Caleb couldn't believe their luck. Now they had a real excuse to go over to Mr. González's place.

As he and Tyson got closer to apartment 14B, Caleb started to have doubts. Maybe Mr. González was in bed with a cold. Perhaps he was just tired of reading newspapers. This could be a silly idea.

"This is a silly idea," he said aloud.

Tyson ignored him and kept walking.

Caleb thought about it again. What if something terrible had happened and Mr. González needed their help? What if it involved a nasty villain? What would Detective Magnet do?

Caleb approached the front door slowly. He was getting nervous. If Tyson wasn't with him, he could have changed his mind. But Tyson was here. Caleb would never live it down if he gave up now.

Just then, a black cat jumped out at them from nowhere.

Chapter 4

Going, Going . . . González!

"ARRGGHH!" screamed Caleb.

"Cute!" said Tyson. He stroked the cat that was now rubbing up against his legs.

"Leroy, you scared me!" said Caleb.

Leroy was Mr. González's black cat. He was never this friendly. It was usually better to stay away from him if you didn't want to get scratched.

"How could you be scared of such a cute kitten?" asked Tyson.

"It's a cat, not a kitten," mumbled Caleb angrily. He was feeling a little embarrassed about screaming.

"I think he's lonely," said Tyson.

Caleb looked at the cat. Something was definitely not right here. First, there were the unopened newspapers and mail, and now Leroy was just wandering the building. Mr. González never allowed Leroy to wander around scaring people.

"Let's knock on the door," Caleb said.

Caleb and Tyson faced the door together. Caleb knocked quietly. Even as he knocked more loudly a second time, he knew no one was home. He wasn't sure how he knew. But there was a stillness about the door, like it was holding its breath.

Caleb let out his own breath. He raised his fist to knock again, and that was when he noticed something else. There was a mail slot in the door with a brass plate that read "Letters" above it. Some of the older front doors still had these slots. They were used in the days when a concierge delivered the mail to your door.

Now the mail slots were used only for messages between the apartment dwellers. Mr. González was president of the residents' association, so he received lots of mail from his neighbors. Caleb could hardly read the sign above the slot because there were so many notes sticking out of it.

Caleb pointed to the mail and Tyson nodded. Then, from somewhere inside the apartment, a phone rang. The boys listened, waiting for footsteps to announce that Mr. González was home. But there were no footsteps. The phone finally stopped ringing. The only sounds left were the demanding cries of Leroy the cat.

"I don't think anyone's home," said Caleb.

Tyson nodded. "Let's get this cat out of the hallway," he said, picking up Leroy. "*Detective Magnet*'s on soon."

But now Caleb was very curious. "Let's just look around a bit," he said.

"What are we trying to find?" asked Tyson.

"Clues," said Caleb. "Come on. Let's see if Mr. González's car is in its parking space in the basement lot."

On their way downstairs, Tyson asked Caleb whether anyone else might have noticed that Mr. González was not at home.

"What does Mr. González do?" Tyson asked.

"He's retired," replied Caleb.

"Okay," said Tyson. "So we can't check out his coworkers. Does he live with anybody?"

"No. Just the cat," Caleb said, watching Tyson rub Leroy's ears.

"Okay, so that's a dead end. What about family?"

Caleb had to admit he didn't know much about Mr. González's family. "I just know he has a dark-blue car. He polishes it every Saturday. He loves that car."

They came out of the elevator into the garage. Caleb pointed toward Mr. González's parking space. "It's over there," he said, "next to ours."

Caleb was leading the way to where his dad parked his car when he saw it. Mr. González's shiny, dark-blue car was parked next to his dad's car. He stared at Tyson.

"This is getting serious!" Caleb said.

Chapter 5

Brainstorming

FIVE MINUTES LATER, Caleb, Tyson, and Leroy were sitting in Caleb's living room. They were watching the last half of *Detective Magnet*. During the show, the private eye made a list of clues to help him solve the mystery of the missing diamond.

"That's what we need to do," said Tyson.

Caleb agreed. At the end of the show, he got a pad and pen from the kitchen. Then he started to make a list.

List of clues:

- two days of newspapers in lobby mailbox
- a pile of mail in lobby mailbox
- a pile of neighbors' notes in front-door mail slot
- no one answering the front door
- the sound of an unanswered phone from inside the apartment
- a lonely cat
- a car still in its parking space

"Now what?" asked Caleb.

"Let's brainstorm. What could this list prove?" asked Tyson.

Caleb made another list.

What the list of clues might prove:

- Mr. González forgot to cancel his newspapers.
- Mr. González forgot to pick up his mail.
- Mr. González does not want to read any more notes from his neighbors.
- Mr. González is feeling tired and he's asleep in bed.
- Mr. González is feeling sick and he can't answer the phone.
- The cat is just pretending to be lonely.
- Mr. González's car doesn't work anymore so he can't use it.
- A nasty toy maker is trying to take over the world.

"What's that last thing you wrote?" asked Tyson.

"Huh?" said Caleb. "Oh, ha ha. Just joking." Caleb scribbled out his last entry.

"Caleb, what if Mr. González really is sick? What if he has hit his head?" said Tyson.

"Maybe we should talk to my dad," agreed Caleb.

Caleb and Tyson took their lists to the kitchen where Mr. Piper was making dinner. "You kids really get into your work, don't you?" he said, smiling.

Caleb asked his father to read the lists. Mr. Piper seemed impressed. "Well done, boys," he said. "You're nothing if not thorough. Look, dinner's ready, so let's eat it first, okay? Then, after *Who's the Smartest?*, we'll talk."

Caleb nodded. He knew his dad didn't like to miss an episode of his favorite game show.

"But what if Mr. González has passed out on the floor?" Caleb whispered to Tyson.

"What if he just didn't want to answer the door?" said Tyson. "Anyway, I've got to go home for dinner. Can you call me later?"

Caleb promised he would. Then he set the table and fed the cat. Leroy ate a whole can of tuna in thirty-two seconds. During dinner, Caleb scarfed down his own food. He cleared the table, even before his father had finished.

Mr. Piper went to the living room and turned on the television. Caleb was too jumpy to sit still and watch. He finished washing all the dishes. He was putting a saucer of water out for Leroy when his dad called out, "Caleb! Quick! Come here!" Caleb rushed into the living room. His father was pointing to the television.

"Look," said Mr. Piper.

There, on the television, were three people standing behind podiums with their hands on buzzers.

"You know I don't like this show, Dad," said Caleb.

"In the middle," said Mr. Piper, pointing. "See who it is."

Just then, the camera zoomed in on the person in the middle.

"The answer is Brazil," said the person.

"Mr. González!" yelled Caleb.

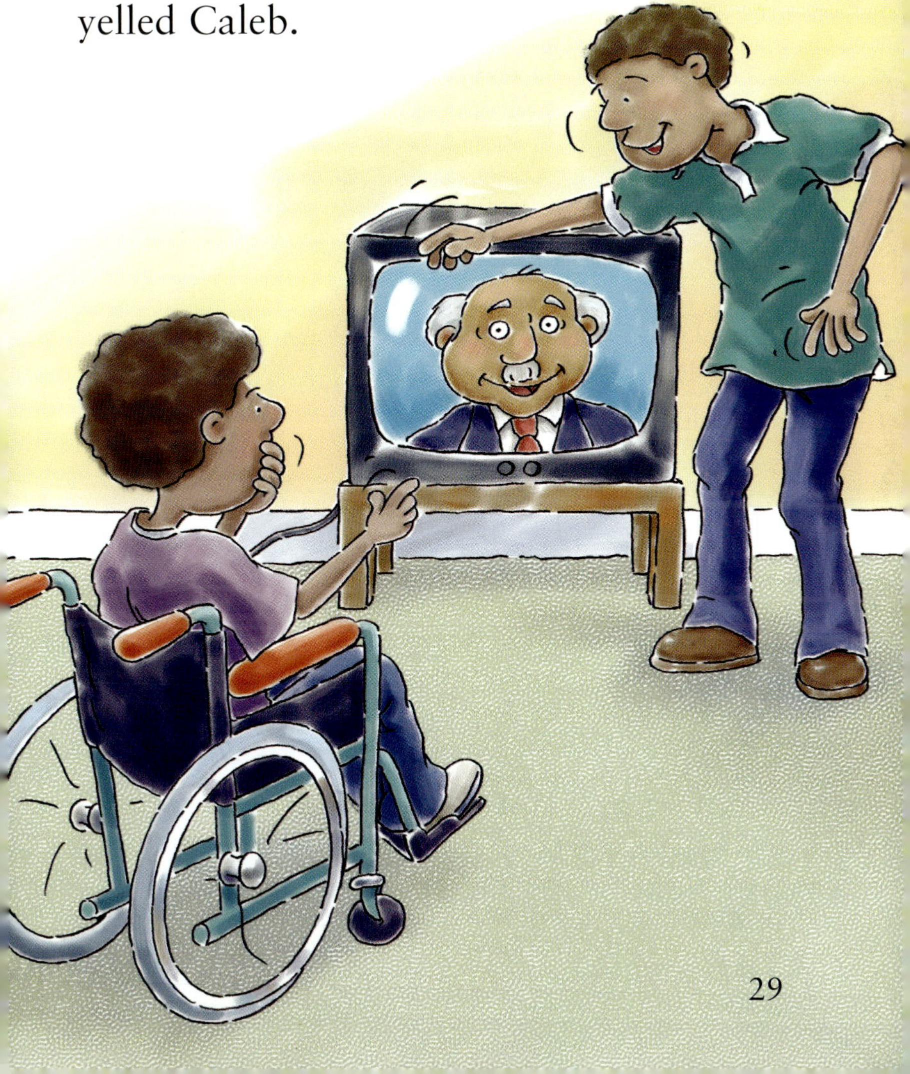

Chapter 6

Mystery Solved?

THE NEXT DAY was Saturday. Caleb and Tyson were working on a detective kit when Mr. González knocked on Caleb's front door.

"I got your note," said Mr. González. "Thank you very much for looking after my cat."

"That's okay, Mr. González," said Caleb. "We saw you on television. Did you win anything?"

"I won a nice car polishing kit, actually," replied Mr. González.

Mr. González then explained that a contestant on *Who's the Smartest?* had become ill at the last minute. Mr. González was next on the list.

"I was supposed to appear next week," explained Mr. González. "It was such craziness when they called me. They film the show live in Los Angeles, you know. I took a taxi to the airport. I completely forgot to cancel my newspapers. But at least I planned for Leroy to be looked after. Well, I thought I had."

Mr. González had asked his next-door neighbor to look after Leroy. When Mr. González returned home, he'd gone straight over to check on his cat. But Leroy wasn't there. The neighbor wasn't there either. A basket of laundry had been left outside her front door, which was slightly ajar.

"It's very strange," said Mr. González. "I just can't imagine what's happened to Mrs. Triffiletti."

Caleb grabbed a pen and paper.

"Could I ask you a few questions, Mr. González?" asked Caleb.

"What are you thinking?" asked Mr. González.

"Mr. González," said Caleb. "I'm thinking we have a mystery on our hands."